Inside the Military

MILITARY WEAPONS

Martha London

DiscoverRoo
An Imprint of Pop!
popbooksonline.com

abdobooks.com

Published by Pop!, a division of ABDO, PO Box 398166, Minneapolis, Minnesota 55439.

Printed in the United States of America, North Mankato, Minnesota.

052019
092019

THIS BOOK CONTAINS RECYCLED MATERIALS

Cover Photo: US Department of Defense

Interior Photos: US Department of Defense, 1, 8–9, 22; Goran Tomasevic/Reuters/Newscom, 5; AB Forces News Collection/Alamy, 6; Alaa Al-Marjani/Reuters/Newscom, 7; iStockphoto, 11, 13, 29; Shutterstock Images, 12, 14, 15, 17 (early gun), 17 (rifle), 17 (bullets), 17 (gun barrels), 18 (top); Library of Congress, 16, 19 (top), 30; Hulton Archive/Getty Images, 18 (bottom); The Protected Art Archive/Alamy, 19 (bottom); Nevada National Security Site, 21; US Air Force, 23; Defense Visual Information Distribution Service, 24, 25, 28, 31; Britta Pedersen/picture-alliance/dpa/AP Images, 27

Editor: Connor Stratton
Series Designer: Jake Slavik

Library of Congress Control Number: 2018964862

Publisher's Cataloging-in-Publication Data

Names: London, Martha, author.

Title: Military weapons / by Martha London.

Description: Minneapolis, Minnesota : Pop!, 2020 | Series: Inside the military | Includes online resources and index.

Identifiers: ISBN 9781532163890 (lib. bdg.) | ISBN 9781644940624 (pbk.) | ISBN 9781532165337 (ebook)

Subjects: LCSH: Military weapons--Juvenile literature. | Armaments--Juvenile literature. | Munitions--Juvenile literature.

Classification: DDC 623.4--dc23

WELCOME TO DiscoverRoo!

Pop open this book and you'll find QR codes loaded with information, so you can learn even more!

Scan this code* and others like it while you read, or visit the website below to make this book pop!

popbooksonline.com/military-weapons

*Scanning QR codes requires a web-enabled smart device with a QR code reader app and a camera.

TABLE OF CONTENTS

CHAPTER 1

MILITARY WEAPONS IN ACTION

Soldiers moved slowly through the streets of Mosul, Iraq. Enemies had taken over the city. US and Iraqi soldiers worked together. They held their guns up so they would be ready to shoot.

WATCH A VIDEO HERE!

US and Iraqi soldiers fought for nine months to take back Mosul, Iraq.

Enemy shooters were hiding high up in buildings. Their guns could shoot from far away. Soldiers used strong guns to shoot these enemies. US planes also dropped bombs over

Iraqi soldiers practice firing M16 rifles. Many militaries around the world use this rifle.

From October 2016 to June 2017, US planes completed 69 bomb-dropping missions over Mosul, Iraq.

the city. These bombs destroyed enemy hiding places. However, they also killed many **civilians**.

DID YOU KNOW?

The M16 rifle can fire 200 bullets per minute.

Bombs and guns are common military weapons. Soldiers use other weapons too. For example, missiles

Soldiers fire a Stinger missile at an aircraft. A Stinger can follow an aircraft even as it moves.

explode like bombs. But they can fly. Soldiers can also guide missiles in different directions.

CHAPTER 2

HISTORY OF MILITARY WEAPONS

Many ancient armies fought with spears. At first, spears' sharp tips were made of stone. But approximately 5,000 years ago, people began working with metal. They made metal spear tips and swords.

LEARN MORE HERE!

Approximately 2,000 years ago, soldiers in the Roman army fought with metal spears.

DID YOU KNOW?

The earliest spears are more than 400,000 years old.

Ancient armies used other weapons too. Longbows could shoot arrows from far away. And **catapults** flung rocks into walls. In the 800s CE, Chinese scientists invented gunpowder. This black powder was explosive. Early bombs used the powder.

DID YOU KNOW?

People first used gunpowder to make fireworks.

Ancient Greeks built the first catapult in approximately 400 BCE.

In the 1300s, armies used gunpowder to build guns. At first, a shooter burned a **wick** attached to the gun. That wick

A man shows how to fire an early gun using a wick.

In the early 1500s, guns had metal wheels. The wheels spun against stones to make sparks. These guns did not need wicks. They could shoot faster than earlier guns.

made a spark. Then, the spark set off the gunpowder. Finally, the explosion shot the bullet. These early guns were slow. They were also hard to aim.

Over time, guns became faster. Guns also got better at aiming. For example, rifles became common in the 1800s. Bullets spin as they leave a rifle. This spin helps bullets hit their targets.

During the 1800s, the United States took over a large amount of American Indian land. These members of the Apache tribe fought with rifles to keep their land.

HOW RIFLES WORK

The barrel is the part of the gun that a bullet travels through. Early gun barrels were smooth on the inside. Rifle barrels have grooves in them. These grooves cause bullets to spin as they leave. This spin makes it easier to aim the bullets.

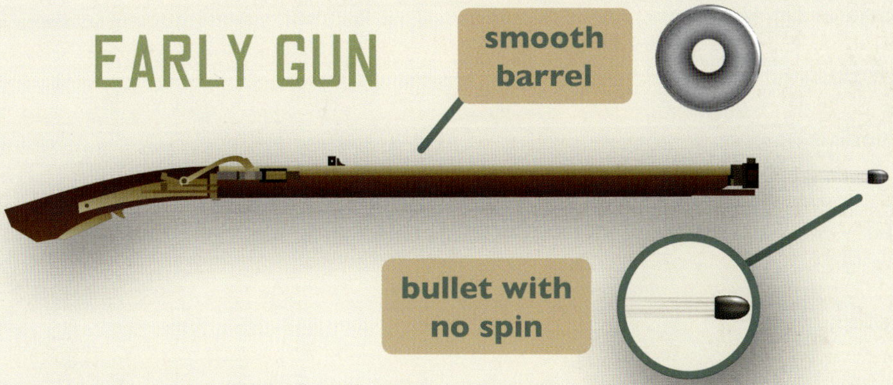

MILITARY WEAPONS TIMELINE

3000s BCE
People begin making weapons out of metal.

800s CE
Chinese scientists invent gunpowder.

1300s
People use guns for the first time.

1945
The United States drops two **nuclear bombs** on cities in Japan.

2002
A laser shoots down a missile for the first time.

1864
US soldiers use the first machine gun.

CHAPTER 3

MILITARY WEAPONS TODAY

Today, military weapons are powerful.

Some countries have **nuclear bombs**.

When these bombs explode, they

send out large amounts of energy.

One nuclear bomb can destroy a city.

COMPLETE AN ACTIVITY HERE!

In 2019, there were more than 14,000 nuclear bombs in the world. Together, the United States and Russia had more than 13,000 of those bombs.

Military weapons are also more **accurate** than before. For instance, some missiles use sensors to find targets

A missile that can track its target helps the soldier stay safe. The soldier can take cover as soon as he or she fires the missile.

One type of missile can track a target nearly 15 miles away.

on their own. Certain sensors can track down heat. Planes make a lot of heat. Missiles can use heat sensors to hit those planes.

Today's weapons can fire much faster than older weapons. For example, many soldiers use machine guns. Firing a bullet creates a lot of force. Machine guns use this force to load another bullet. This system lets machine guns fire many bullets quickly.

One common machine gun in the US military can fire 800 rounds per minute.

A soldier helps feed bullets into a machine gun.

DID YOU KNOW?

In 2007, a company built a gun that fires one million rounds per minute.

CHAPTER 4

THE FUTURE OF MILITARY WEAPONS

Scientists continue to build stronger guns. The 338TP is one example. This gun uses sensors like missiles do. It can lock on to targets. The 338TP can hit moving targets from nearly a mile away.

LEARN MORE HERE!

The 338TP can lock on to targets moving as fast as 20 miles per hour.

Militaries have also begun using lasers as weapons. A laser is a narrow ray of light. Lasers heat the objects that

Most long-range weapons can run out of their bullets or missiles. A laser can fire as long as it has energy.

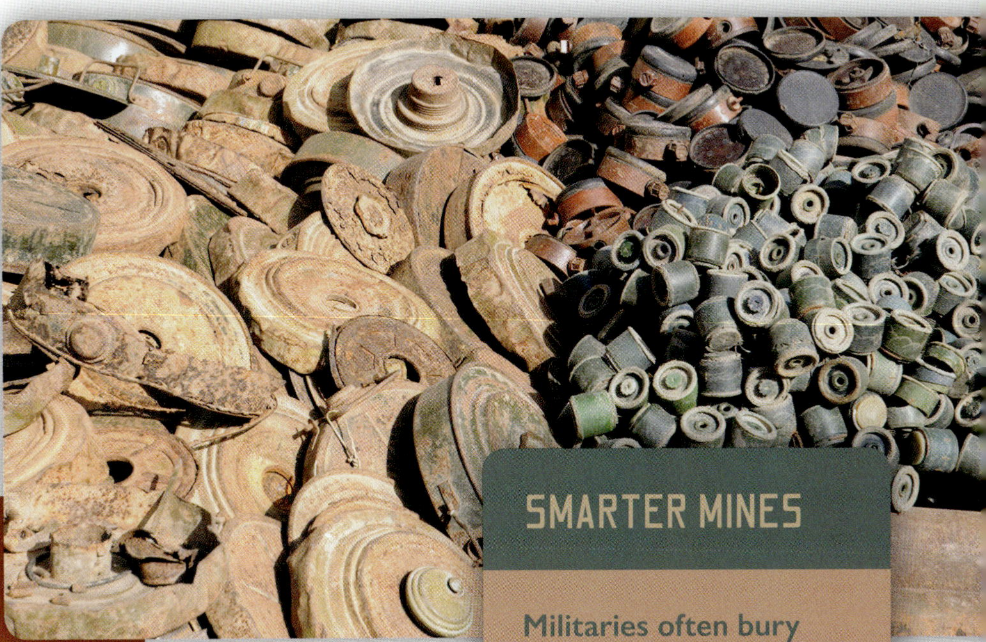

SMARTER MINES

Militaries often bury **mines** below ground. If someone steps on a mine, it will blow up. Mines can kill **civilians**. The US military is trying to build safer mines. These mines use sensors to notice when someone is nearby. If the person is not an enemy, soldiers can stop the mine from blowing up.

they hit. If a laser is strong, its heat can be harmful. Lasers are cheaper than missiles. But they hit targets just as well.

MAKING CONNECTIONS

TEXT-TO-SELF

Which military weapon do you think is the best? Why?

TEXT-TO-TEXT

Have you read books about other military tools? How are they similar to military weapons? How are they different?

TEXT-TO-WORLD

Military weapons are powerful. What should soldiers think about before using them?

GLOSSARY

accurate – able to hit the intended target.

catapult – a weapon that throws large stones through the air.

civilian – a person who is not part of the military.

mine – an explosive weapon often put underground.

nuclear bomb – a strong explosive that works by breaking atoms apart.

wick – a piece of woven fiber that can be burned in candles, lamps, or early guns.

INDEX

ONLINE RESOURCES

popbooksonline.com

Scan this code* and others like it while you read, or visit the website below to make this book pop!

popbooksonline.com/military-weapons

*Scanning QR codes requires a web-enabled smart device with a QR code reader app and a camera.